Mario Sergio Frankievicz

Geotechnologies applied to Higher Education in Geography

Mario Sergio Frankievicz

Geotechnologies applied to Higher Education in Geography

Geotechnologies and Geography Teaching

ScienciaScripts

Imprint

Any brand names and product names mentioned in this book are subject to trademark, brand or patent protection and are trademarks or registered trademarks of their respective holders. The use of brand names, product names, common names, trade names, product descriptions etc. even without a particular marking in this work is in no way to be construed to mean that such names may be regarded as unrestricted in respect of trademark and brand protection legislation and could thus be used by anyone.

Cover image: Provided by the author

This book is a translation from the original published under ISBN 978-613-9-71731-6.

Publisher:
Sciencia Scripts
is a trademark of
Dodo Books Indian Ocean Ltd. and OmniScriptum S.R.L publishing group

120 High Road, East Finchley, London, N2 9ED, United Kingdom
Str. Armeneasca 28/1, office 1, Chisinau MD-2012, Republic of Moldova, Europe
Printed at: see last page
ISBN: 978-620-7-66275-3

ACKNOWLEDGEMENTS

To God, in the first place.

To my supervisor, Siumara Aparecida de Lima, for her attention and dedication.

To my co-supervisor Kleber Rodrigo Penteado, for his attention and dedication.

UTFPR, for the opportunity to attend the specialisation course in Scientific and Technological Education.

To the teachers who worked for us over the course of a year, for their friendship and encouragement.

To our colleagues, for the time we shared.

To all the people who helped in the preparation of this work.

SUMMARY

FRANKIEVICZ, Mário Sérgio. **Geotechnologies applied to higher education in Geography at an educational institution in the city of Ponta Grossa.** 2010. 34 f. Monograph (Specialisation in Scientific and Technological Education) - Federal Technological University of Paraná. Ponta Grossa, 2010.

The research looked at what current geotechnological resources can be used in higher education in Geography, how geotechnologies are used in teaching practice and how teachers assess the use of geotechnologies in their students' education. During the research, the methodology applied was through semi-structured open questionnaires carried out with higher education Geography teachers. The first chapter discusses some theoretical considerations about the historical development of geotechnologies in Geographical Science and how these resources are used in higher education in Geography. In the following chapter, the theoretical considerations addressed the training of teachers, geographers and how teachers have been using geotechnologies in higher education in Geography. Based on this panorama, we analysed how geotechnologies can contribute to the education of undergraduates and/or graduates in Geography. In the last chapter, we explored the interviews in order to demonstrate the vision of Geography teachers in relation to the use of geotechnologies in teacher education. The final considerations point to some issues that could be improved through the development of new research into the application of geotechnology in the training of geography teachers.

Keywords: Geotechnology. Geography teaching. Higher education.

SUMMARY

1 INTRODUCTION

Geotechnologies have contributed to the development of many scientific works, especially in the exact and natural sciences, such as Geography, Geology, Biology, Physics, Chemistry and Engineering such as Cartography, Environment and Forestry.

According to Fitz (2008, p. 11). "The inclusion of professionals from different areas of knowledge, especially geographers, is essential for the successful outcome of the work carried out."

In order to do this, it will be necessary to emphasise, according to Fitz (2008, p. 13):

> The close link between geotechnologies and scientific concepts related to geographical science. This condition will lead to an understanding of the processes, procedures, analyses, etc. linked to Geographic Information Systems (GIS) and geoprocessing techniques.

Based on this premise, we realise that Geography must evolve together with geotechnology, so that Geotechnology becomes a more present and effective tool in the field of Geographical Science. At least in Brazil, the importance given to geotechnology is still small. There aren't many scientific events and, above all, a large number of professionals working in the field of Geography are deficient in computer skills.

According to Fitz (2008, p. 14), Geographical Science has been evolving gradually, while technological evolution has been accelerating. As a result, Geography is increasingly adapting to scientific and technological advances.

For this reason, this paper raises the issue of "how geotechnologies are used in teaching practice on a degree course in Geography". It deals with what current geotechnological resources can be used in higher education in Geography, how geotechnologies are used in teaching practice and how teachers evaluate the use of geotechnologies in the training of their students in higher education in Geography.

To this end, the objectives set for the work are:

General objective:

- To analyse the contributions of geotechnologies to higher education in geography.

Specific objectives:

- To identify which geotechnological tools are used in the undergraduate degree course in Geography;

- To diagnose how geotechnological resources are used by undergraduate geography teachers and how they are included in their teaching practice;

- Characterise how the coordination assesses the use of Geotechnologies in the training of its student body.

This research is justified by the fact that as an academic I felt the need to have a better grounding in geotechnologies and their applications for teaching geography. During my undergraduate studies, I fell short in some activities that required the use of geotechnological knowledge, such as the Compulsory Supervised Internship in Geography and the Scientific Initiation programme.

During the research, the methodology applied was through semi-structured open questionnaires carried out with professors working in higher education in Geography, for which the professors were divided into two categories. The first is the professors who teach subjects directly or indirectly linked to geotechnologies, identified as P1, P2, P3 and P4, and the second category is the course coordinator, identified in this paper as PC. A digital camera with audio and video was used to film the interviewees.

The first chapter is the introduction to the work. Chapter two discusses some theoretical considerations on the historical development of geotechnologies in Geographical Science and how these resources are used in higher education in Geography.

In chapter three, theoretical considerations will focus on teacher training, geographers and how teachers have been using geotechnologies in higher education in Geography. Based on this panorama, we will analyse how geotechnologies can contribute to the education of undergraduate and/or graduate students in Geography.

In chapter four, the interviews will be explored in order to demonstrate the view of geography teachers regarding the use of geotechnologies in teacher training.

Finally, the concluding remarks point to some issues that could be improved through the development of new research guided by the concern for teaching that is concerned with the application of geotechnologies in the training of geography teachers.

2 A DISCUSSION ON GEOTECHNOLOGIES AND TEACHING HIGHER EDUCATION IN GEOGRAPHY

This chapter deals with some theoretical considerations on the historical development of geotechnologies within Geographical Science and the approach of these resources in higher education in Geography.

2.1 THE HISTORICAL ORIGINS OF THE USE AND DISCUSSION OF GEOTECHNOLOGIES

Throughout the historical development of geographic science, various currents of thought have emerged. One of these currents emerged in the 1970s: Theoretical-quantitative geography, which discusses issues related to statistical surveys and also the capture and tabulation of data related to geotechnologies.

> For authors affiliated with this current, geography could be explained entirely using mathematical methods. All the issues dealt with there - the relationships and interrelationships of phenomena and elements, local landscape variations, the action of nature on men, etc. - could be expressed in numerical terms (by measuring their manifestations) and understood in the form of calculations. Advances in statistics and computing provide a geographical explanation (MORAIS, 1995, p. 102).

Based on this reality, it is important to emphasise that geotechnologies are of fundamental importance in the geographer's and geography teacher's approach to collecting and tabulating data relating to the study of space.

With this in mind, this paper proposes a reflection on the geotechnologies used in a university's Bachelor's and Bachelor's programmes between 2004 and 2009.

2.2 GEOTECHNOLOGICAL RESOURCES APPLIED TO THE HIGHER EDUCATION OF GEOGRAPHY

In order to continue our discussion, it is necessary to conceptualise some geotechnology tools, among which we will highlight GIS (Geographic Information Systems), Remote Sensing and

GPS (Global Positioning System).

In order to conceptualise the first geotechnology we are going to discuss, it is worth using the idea of FITZ (2008), who explains that "the acronym GIS stands for Geographic Information System. It is a computer system that works with an infinite amount of geographic information". According to the same author, GIS can be defined as:

> A system consisting of a set of computer programs, which integrates data, equipment and people in order to collect, store, retrieve, manipulate, visualise and analyse data spatially referenced to a known coordinate system (FITZ, 2008, p. 23).

In higher education in geography, GIS can be used to build cartograms or other cartographic documents, giving geography students a more applied and accurate use of cartography in their practice. In the professional field, geographers can use GIS for scientific research, teaching or working in a wide variety of public and private organisations.

According to the teachers interviewed, geotechnologies are used in geography teaching with the following GIS tools: Spring, Gvsig and Quantum Giz (free software) and Arcview (proprietary software).

These Geographic Information Systems can be used in higher education in Geography in the following activities according to Fitz (2008, p. 26)

> ° Updated mapping of the municipality;
> ° Various zonings (environmental, socio-economic);
> ° Monitoring risk areas and environmental protection;
> ° Structuring energy, water and sewage networks;
> ° Tax tariff adjustment;
> ° Urban expansion studies and modelling;
> ° Control of irregular occupation and construction;
> ° Establishment and/or adaptation of transport modes; e,
>
> ° Spatial analyses using various thematic maps, among others.

Another geotechnological tool of great relevance to both geography teaching and the geographer's profession is remote sensing, which in turn can be understood as follows

> Broadly speaking, it's a way of obtaining information about an object or target without making physical contact with it. The information is obtained using electromagnetic radiation reflected and/or emitted by targets, generated by natural sources such as the Sun and the Earth, or by artificial sources such as Radar (ROSA, 2005, p. 83).

In this sense, the main characteristic of remote sensing is that it obtains information about objects without the proximity of any individual.

> Remote sensing basically involves two phases: the data acquisition phase and the utilisation phase. In the acquisition phase, information is provided on electromagnetic radiation, sensor systems, the spectral behaviour of targets, the atmosphere, etc? The utilisation phase mentions the different possibilities for applying this data in various areas of knowledge, such as Geography, Agronomy, Civil Engineering, Geology, Hydrology, Pedology, etc. (ROSA, 2005, p. 83).

The two basic phases of Remote Sensing make it possible to better utilise the resources obtained from satellite images in various branches of science, especially in higher education in Geography, because with Remote Sensing techniques it is possible to develop various types of work, both in the field of scientific research by teachers and academics, and in teaching practice, as we will see below.

In the field of research, remote sensing will make it easier for researchers to quickly acquire data and then use it in their work,

> . however, their potential in geographical studies has not been sufficiently exploited. This is largely due to deficiencies in initial training and the lack of continuing training for many professionals, which is essential in order to keep up with growing technological advances (FLORENZANO, 2005, p. 24).

In teaching practice, they will provide teachers with an important subsidy in the production of didactic material, which will help them transpose knowledge to their students.

The interpretation of satellite images, combined with geographical concepts and field practice, provides students with a considerable amount of information to the point where they are able to carry out a more in-depth analysis of geographical space.

In universities, remote sensing is used in practical and theoretical classes. In addition to GIS and remote sensing, another geotechnology used in higher education in Geography is GPS1.

The Global Positioning System (GPS) is of fundamental importance because a geotechnological resource such as this can be included in the pedagogical practice of a teacher linked to the teaching of Geography because it offers a wide range of possibilities, facilitating the transmission of geographical knowledge to the student. In the teaching practice of a geography professional, GPS can be used on field trips, with the aim of mapping a micro or meso space by means of marked points and, from these, constructing a digital cartographic document whose purpose will be to characterise the space travelled with the GPS device. The student will have the experience of operating the GPS on a field trip, thus gaining knowledge of a new geotechnological tool that can make it easier to build a map.

If we take into account the number of students enrolled on the BA and BSc courses[1,2] , we can see that there is a basic structure to cater for its public.

This chapter presents some of the geotechnologies used in higher education in Geography and which can be used by any university that offers a degree or bachelor's degree in Geography.

---

[1] GPS is a satellite-based radio navigation system developed and controlled by the United States Department of Defence (U.S. DoD) that allows any user to know their location, speed and time, 24 hours a day, under any weather conditions and anywhere on the globe (ROSA, 2005, p. 85).
[2] An average of 40 students per class, according to information provided by the course coordinator.

3 GEOTECHNOLOGIES APPLIED TO THE TEACHING PRACTICE OF GEOGRAPHY PROFESSIONAL

This chapter deals with some theoretical considerations about teacher training, geographers and how teachers have been using geotechnologies in higher education in Geography, outlining how these geotechnologies can contribute to the training of undergraduate and/or graduate students in Geography.

3.1 THE TRAINING OF TEACHER-RESEARCHERS IN GEOGRAPHY TEACHING

In the historical development of Geographical Science, there has been much discussion about the role of the Geography teacher. Whether it's the teacher who simply transposes the content to the student or whether, together with the first question, he or she really assumes the position of a research teacher.

> The issues surrounding academic research and teacher training was a proposal defended by lay education, which fundamentally proposed an education system suited to political and economic needs (LIMA, 2004, p. 121).

In other words, lay education can be seen here as basic education that is based solely on the interests of the state and not on the academic training required to become a research professor, but rather on the specialisation of the workforce, which must necessarily go through school (LIMA, 2004, p.121).

According to Lima (2004, p. 121), the structure of public universities would be dedicated to scientific research, while private universities would offer university courses that required little infrastructure. Thus, degrees such as bachelor's degrees would be offered in private colleges, but with the premise that future academics would be trained to reproduce existing knowledge and not to build up a new theoretical framework gradually through individual or group scientific research.

Research will contribute to geography teaching when it also takes into account the concrete reality found in the Brazilian education system (LIMA, 2004). In general, academic research will only have a significant impact on teacher training if it takes into account the importance of didactic knowledge (LIMA 2004, p. 124). With didactic knowledge, it is possible to build a more solid education that is coherent with the professional practice of geography graduates.

This is why geography teachers must produce their own scientific knowledge and not merely reproduce it.

3.2 THE PROBLEMS LINKED TO THE TRAINING OF PROFESSIONALS WITH BACHELOR'S AND MASTER'S DEGREES IN GEOGRAPHY

Historically, the practice of geography professionals has not changed considerably, which is why "the vast majority of geographers have worked as geography teachers in primary, secondary and tertiary education" (OLIVEIRA, 1989, p. 28). In other words, from the 1961 Law of Guidelines and Bases to the more recent Law of Guidelines and Bases 9394/96, which guides educational principles today, the geography teacher has not continuously participated in the process of discussing geography.

> What actually happens is that teachers (all of them), obviously geography teachers too, are involved in a dialectical process of domination, whereby teachers are educated to teach without questioning the content of the textbooks, without the end product of their teaching being tools with which they and their students will transform the teaching they do and, of course, the society in which they live. In other words, teachers and students are trained not to think about what they have been taught, but to simply repeat what they have been taught. This means that they don't participate in the process of producing knowledge (OLIVEIRA, 1989, p. 28).

So we can say that geography teachers have not been actively participating in the construction of geographical knowledge and this may be associated with the lack of resources that were absent in their academic training, including those linked to geotechnology. Geotechnological tools could provide them with a wider range of knowledge, thus making their teaching practice more agile and objective.

In this respect, we must also take into account the division of the Geography course into a degree and a bachelor's degree, which does not always end up being fruitful for the subject, as we can see below.

> This division of courses can only contribute to the further impoverishment of Geography and the natural death of Geography teaching, and in particular of the school as a privileged locus for the critical formation of the men who make up and will make up the productive base of society in the future. It's worth pointing out

> that several Geography departments in Brazil have created this mistaken separation of undergraduate programmes, or are opportunistically waiting for the moment of inevitable curricular change to do so. And this opportune moment may be being allowed, among us geographers, by the implementation of the law that regulated our professional practice and which did not allow this practice to Geography graduates. (OLIVEIRA, 1989, p. 29).

The discussion that permeates the current configuration of Geography teaching is still quite debatable, because with the separation of undergraduate courses into a Licence and a Bachelor's degree, there has been a serious dismemberment of vital knowledge for the two categories of course. These include geotechnologies, which are not currently part of the curriculum of some undergraduate Geography programmes at Brazilian universities.

This lack of knowledge about geotechnologies leads to gaps in the training of Geography graduates and undergraduates. For graduates, geotechnology is of fundamental importance even for placement on the labour market, otherwise they obviously won't even get their first job as a geographer. And as far as graduates are concerned, they will gain geotechnological knowledge that will make their lessons more dynamic and allow them to apply their knowledge to the benefit of their students, providing quality lessons and up-to-date knowledge that is closer to reality.

One example that can be cited in the teaching practice of Geography teachers is in relation to the content of the subject of Cartography, which refers to the contents of geotechnology. Many teachers find it difficult to explain the basics of cartography in the classroom, such as definitions of graphic and numerical scales, geographical coordinates and cartographic projections, due to a lack of theoretical and technical support, which has probably not been worked on properly or has not even been incorporated into their academic training.

According to VISENTINI (2004), "there is no point in proposing or realising (as many courses do throughout Brazil) a rigid separation between the bachelor (the geographer) and the graduate (the teacher), as if the latter didn't need scientific training[3] ".

For this reason, it would be interesting for Brazilian universities to review the curricula of Geography courses in the country. According to the National Curriculum Parameters for Higher Education in Geography, universities should include basic subjects related to geotechnology, such as

---

[3] Therefore, the guideline for a geography course that aims to train good professionals (whether teachers or not, it doesn't matter) is to have an adequate basic course: one that is pluralistic and takes in the various areas and trends of geographical science; one that is geared not towards producing specialists but towards developing students' ability to "learn how to learn", to research, to observe, to read and reflect, to be suspicious of clichés or stereotypes, to take initiative and have their own skills. With this - in other words, students who keep up with debates, new themes and new ideas, who are encouraged to think and observe on their own, who acquire a minimum mastery of research techniques, library or archive surveys, etc. - a good professional is being formed who can teach or join a team working in another activity. (VESENTINI, 2004, p. 240)

Cartography and Thematic Cartography, in their curricula for the basic training of Geography professionals, However, subjects such as Remote Sensing, Geographic Information System or Geotechnology applied to Geography are optional, so universities can establish syllabuses that take into account the differences between bachelors and graduates, thus causing the problem we are pointing out related to the training of professionals in the field of Geography, especially graduates. The latter end up suffering from a lack of important technical resources in their academic training, both in the field of research, i.e. the production of knowledge, and in their teaching practice.

3.3 THE USE OF GEOTECHNOLOGIES IN HIGHER EDUCATION IN GEOGRAPHY.

During the course of this work, we looked at aspects relating to the training of a researcher teacher in Geography and what mechanisms would be needed for him to be a good professional, with the ability to produce his own knowledge and not just reproduce it. From now on, we'll change the focus, emphasising how these geography professionals use geotechnology in their pedagogical practice, in other words, what they do inside and outside the classroom as geography teachers.

In the course of the research, a number of professors from a higher education institution were interviewed. They reported on how they use geotechnology tools in the subjects they teach and in the field of research in which their classroom students take part as volunteers or scholarship holders within a scientific initiation programme.

The following reports help us to understand the importance that each education professional is giving to the use of geotechnologies in their teaching practice, as well as which are the main geotechnologies they use.

Teacher P1 states that (oral information)[4] :

> *"... I taught Cartography and Thematic Cartography. Cartography is the geographer's tool par excellence, as Lacoste (1989) said, and is inseparable for both graduate and undergraduate professionals, as it deals with the spatial representation of reality. I used geoprocessing and digital cartography techniques in the subjects I taught. Even though these topics are not compulsory on the syllabus. It would be necessary for every teacher to have a good knowledge of cartography, because it is from cartography that the knowledge of geographical science is analysed."*

---

[4] Interview given on 7 April 2010.

Teacher P2 reports (oral information):[5]

> *"... I make direct use of geotechnology in the subject of Cartography, where I teach classes that contain some geoprocessing topics in both theoretical and practical classes. By choice, I only work with free software such as GVSIG and Quantum Giz. There are ways of working with free software and also free information that is made available by research institutions such as IBGE. I use geoprocessing in the subject of Cartography to train teachers who are able to produce their own cartograms and not just use textbooks to bring knowledge to students and also encourage them to take other actions."*

Teacher P3 said (oral information):[6]

> *"... I work with the subject of Biogeography, which is linked to the area of Geotechnology. I use technologies such as GPS in my lessons, which helps to locate different types of vegetation or ecosystems. I use remote sensing for tree-planting projects or identifying vegetation. I emphasise that I can use free software to develop geotechnology within the Biogeography syllabus, but I prefer proprietary software because it's easier for me to work with."*

Finally, teacher P4 says (oral information)[7] :

> *"... I teach Cartography and Topography, but only for the Bachelor of Geography course. During my lessons I use a number of geotechnological tools. These include surveying techniques and digital cartography. I go on field trips with my students to carry out topographical surveys and digital mapping."*

Based on the reports, we can see that all the teachers interviewed use geotechnological resources in the subjects they teach. We can also say that the majority of teachers use geotechnologies in their Bachelor's degree in Geography, but in their Bachelor's degree in Geography the situation is the opposite, which leads us to return to the question of the need for a reform of the Geography curriculum, as we discussed earlier.

Many students are interested in the field of geotechnology, whether they have a bachelor's degree or a degree in Geography, but we can see that:

> The little prominence given to geotechnologies at geographic events (at least in Brazil) reflects the alienation of a wide range of researchers, who simply disregard this tool as part of doing geography. For them - and there are many - the use of modern technologies by geographers would be restricted to consulting the Internet

---

[5] Interview given on 7 April 2010.
[6] Interview given on 15 December 2009.
[7] Interview given on 20 April 2010.

or even typing up texts. (FITZ, 2008, P. 15)

Undergraduate students don't have as much contact with geotechnology, because subjects relating to geotechnology are not part of their course curriculum, and there is also a noticeable lack of discussions that permeate all subjects with the use of geotechnologies.

The universities that offer a degree in Geography differ in their curricula, as some have at least one subject related to geotechnology, while others don't even include a subject related to this area. It would therefore be interesting to restructure the curricula of some universities in order to correct this flaw and to meet this latent need in the course, providing more consistent academic training, especially for graduates.

Throughout this chapter, we present how geotechnologies can be applied in the teaching practice of Geography professionals and what contributions they could make, especially to Geography graduates. In the next chapter, we'll try to explore the vision of geography teachers in relation to the use of geotechnologies in teacher training.

4 THE USE OF GEOTECHNOLOGIES IN TEACHING PRACTICE

In this chapter we will look at some considerations proposed by teachers for the use of geotechnologies in the teaching training of geography graduates.

4.1 GEOTECHNOLOGY AS A PEDAGOGICAL PRACTICE IN EVERYDAY LIFE TEACHER

Having discussed the training of professionals in the field of Geography and a series of technological devices that can help these professionals in their day-to-day practice in the previous chapters, we will now look at geotechnologies as an integral element of pedagogical practice by analysing interviews kindly provided by teachers at a higher education institution.

During the interviews, the teachers commented on their pedagogical practice combined with the use of geotechnological resources and their importance for Bachelor's and Licentiate's degrees in Geography.

We can see in teacher P1's speech that he is in favour of using geotechnologies to teach geography (Oral information)[8] :

> *"... In my opinion, geotechnology can help a lot in the training of geography professionals, because it has a very wide range of possibilities, among which I can mention Remote Sensing, GIS and GPS devices, which can help in the planning and execution of lessons, both in a theoretical and practical lesson, where field trips can be carried out. "*

For teacher P1, geotechnology can help in the more solid academic training of the geography professional and, as LIMA (2004) says in the previous chapter, it can train a research teacher and not a mere reproducer of knowledge.

Professor P1's comments remind us that the subjects in the area of geotechnology are of fundamental importance for both graduates and undergraduates in Geography. As an example of this, we can mention geotechnology involving Remote Sensing, which helps to display maps and satellite images in real time using simple software, such as Google Earth, which is easy to manipulate and can therefore contribute to the teaching practice of Geography teachers.

In the current labour market, few geography graduates have knowledge of geotechnology

---

[8] Interview given on 7 April 2010.

and this is mostly due to shortcomings in the training of these professionals, as we discussed in the previous chapter, and partly because many are not interested in researching, studying or learning more about the subject. This information became clear from the comments that occasionally came up during the interviews with the teachers.

We can see in teacher P1's speech an example of someone who, despite the lack of content in his training, sought to learn how to work with geotechnologies (Oral information)[9] :

> *"... I've only acquired a good grounding in geotechnology due, for the most part, to my individual efforts and also with some of the teachers I sought out during my degree. I'm generally prepared, but I'm not as up to date with current geotechnologies."*

Professor P1's account helps us to realise that geography professionals should seek to deepen their knowledge of the theory and practice of geotechnology. However, it is known that higher education institutions must take steps to include a greater number of subjects related to the area of geotechnology in their curricula, otherwise the situation will only get worse and new professionals, who graduate year after year, will leave the academy without having the minimum knowledge on the subject, unless they do as Professor P1 did and seek out on their own the knowledge that was previously lacking in their training.

Teacher P2 has the same opinion as the teacher mentioned above regarding the adoption of geotechnology in academic geography teaching (Oral information)[10] :

> *"... The geography teacher should leave university as a researcher and not a simple reproducer of knowledge and for me geotechnologies provide a differentiating factor in the training of geography teachers."*

Professor P2's comment makes it clear that geography teachers in higher education should include geotechnology content in the subjects they teach, either directly or indirectly, in order to contribute to the training of professionals. The inclusion of geotechnology in the higher education institution researched has been evolving over the years and the Bachelor of Geography course has stood out in its use of geotechnological tools.

For teacher P2 (oral information)[11] :

> *"... Geotechnological tools are indispensable for geographers, because academics are extremely dependent on them to stay in the labour market, which has been very restricted*

---

[9] Interview given on 7 April 2010.
[10] Interview given on 7 April 2010.
[11] Interview given on 7 April 2010.

in recent years."

Mastering geotechnological tools can provide students with the opportunity for their first job in the technical areas of: urban planning, environmental analysis, as well as other activities related to the geographer's profession. As well as being important for bachelors, the field of geotechnology is also of great value to graduates, as we can see from the words of teacher P3 (Oral information)[12] :

> *"... subjects related to geotechnology in the degree course leave the geography teacher prepared to carry out a good research project and for concrete teaching practice."*

If we take teacher P3's account into account, we can see that knowledge of the theory of geotechnology can make it extremely easy for geography teachers to carry out their lessons, for example if they take their students to a computer lab in their school to work with educational software related to geotechnology.

INPE (National Institute for Space Research) offers free software such as Spring[13] which can be used to build cartograms, contributing to students' learning in the classroom and, at the same time, giving academics an introduction to scientific research, where they can apply it as a tool.

Through individual actions in their pedagogical practice, geography teachers can contribute to the development of a more solid academic education for their students. As quoted by teacher P3 (oral information)[14] :

> *"... through geoprocessing I work with vegetation mapping and in research projects involving urban afforestation in the Biogeography subject."*

Geography teachers must try to innovate in their teaching techniques in order to provide students with a consistent theoretical foundation.

In the interviews it was possible to see that some teachers do not use geotechnology in their teaching practice, and that this use is limited to a small number of teachers. In addition, it was possible to see that, normally, the degree course has only the subjects of Cartography and Thematic Cartography on its curriculum, which can be considered too little, given the need we mentioned earlier for more subjects linked to the area of geotechnologies.

---

[12] Interview given on 15 December 2009.

[13] This software also has educational virtual simulators that cover geotechnology content, which can be applied in primary and secondary schools.

[14] Interview given on 13 December 2009.

Faced with this situation, we can cite the words of teacher P4, who reminds us of the importance of a curriculum reform involving the addition of at least one more subject to the degree syllabus (Oral information)[15] :

> *"... it will be necessary to implement the subject of Remote Sensing in the curriculum of the degree course in Geography, because Remote Sensing will provide a much broader view of geographical space and how it can be represented in countless ways".*

On the other hand, the Bachelor's degree in Geography has a more complete syllabus, including some subjects linked to the area of geotechnology, such as: SiGs (Geographic Information Systems), Remote Sensing, Topography, Cartography and Thematic Cartography. We can see in Professor P4's speech the importance of these subjects in the training of geographers (Oral information)[16] :

> *"... I believe that the Geotechnologies subjects are fundamental for the exercise of the profession and for an opportunity in the labour market, because without geoprocessing there is no way to carry out good work both in the field of research and in their professional practice."*

Professional geographers must take into account that geotechnology is a fundamental tool for developing their practice, as there is great competition in the labour market with professionals from other areas such as engineering and architecture, and in this case, knowledge and good use of geotechnological resources can be the difference for professionals in the field of geography. This, combined with good relationships and clarity in the presentation of ideas, can make the candidate for a job position surpass their competitors.

From what has been discussed, we can see that the teaching of Geography must evolve in line with the development of geotechnologies so that geotechnology can become a more present and effective tool in Bachelor's and Degree courses in Geography.

4.2 THE COURSE COORDINATORS' VIEW OF THE CURRICULUM STRUCTURE

This last part of the paper discusses the problems of the curriculum structure of degree and bachelor's programmes in Geography from the point of view of a course coordinator.

---

[15] Interview given on 15 December 2009.
[16] Interview given to the author on 20/04/2010

According to PC, in 2004, when the current Geography degree programme was set up, there wasn't enough physical space to implement a curriculum that included at least one subject linked to geotechnology (oral information)[17] :

> *"... The installation of a Geoprocessing laboratory would be very important for the Geography programme, but the lack of equipment was too great to set up a laboratory. The number of students taking the course is very high, around 44 per year, which makes it difficult to include subjects such as GIS and Remote Sensing, because these subjects use laboratories where there can't be a large number of students."*

The problem of the lack of laboratories only began to be solved at the institution where PC works in 2007, when the Geography programmes were given a basic geoprocessing research laboratory and a teaching computer laboratory. According to PC (Oral information)[18] :

> *"... Even though there are no subjects linked to geotechnology on the degree course, in the last three years there have been teachers who have worked on some geoprocessing techniques in practical classes within the cartography subject. Teachers who had technical mastery of geoprocessing tools. Also, during the formation of the first cohort of the new programme, there weren't enough effective Geotechnology teachers on the teaching staff of the institution."*

The last part of the quote from the coordinating teacher raises an issue that is not unique to Geography courses in Brazil. It is the reality of many public educational institutions in the country, where there is a shortage of permanent teachers in a wide variety of areas, resulting in a large number of teachers being hired on a temporary basis. These temporary contracts end up interrupting the sequence of work that has been started and can often jeopardise the development of some areas, according to the information provided during the interviews.

Although there are curricula with structural problems in the syllabuses of the degree and bachelor's programmes, there is also the prospect of improvement and curricular reform in the future, according to what was reported in the interviews (Oral information)[19] :

> *"... There is great interest on the part of teachers in implementing at least one subject linked to geotechnology for the next reform of the curriculum, and the Geography course (today) has sufficient structure for this to happen"*

It is hoped that in future Geography degree courses will include at least one subject linked

---

[17] Interview given to the author on 20/04/2010
[18] Interview given on 13 April 2009.
[19] Interview given on 13 April 2009.

to the area of geotechnology, as this will have a major influence on the graduate's academic training, thus providing them with a more complete education in two areas: in their teaching practice and in their development as a research teacher.

In the course of this chapter, we present the use of geotechnologies in the pedagogical practice of geography teachers, as well as the vision of a course coordinator in relation to the composition of the curriculum, especially the bachelor's degree, while also discussing the possibility of restructuring the curriculum.

5 FINAL CONSIDERATIONS

A number of objectives were proposed during the research. These include *Identifying the geotechnological tools used in the undergraduate degree course in Geography at a higher education institution*. The university surveyed uses a number of geotechnological resources, including remote sensing, GIS (Geographic Information Systems) and GPS (Global Positioning System) devices.

Another objective was to *diagnose how geotechnological resources are used in Geography undergraduate courses by teachers and how they are included in their teaching practice*. Based on the teachers' reports, it can be seen that some teachers use geotechnological resources more in the subjects they teach and others use them less. We can also say that the majority of teachers make use of geotechnologies in the Bachelor of Geography course, but in the Bachelor of Geography the situation is the opposite, which leads us to return to the question of the need for a curriculum reform for the Geography course.

To this end, it would be interesting to restructure the curriculum of the institution's course in order to correct this flaw and meet this latent need in the course, providing more consistent academic training, especially for graduates.

We also looked at *how teachers assess the use of geotechnologies in the education of their students*. Geography professionals should endeavour to deepen their knowledge of the theory and practice of geotechnology. However, it is known that higher education institutions must take steps to include a greater number of subjects related to geotechnology in their curricula. Otherwise, the situation will only get worse and the new professionals, who graduate year after year, will leave the academy without the minimum knowledge on the subject.

Geotechnological tools can provide geographers with the opportunity for a first job. In urban planning techniques, environmental analysis, as well as other activities related to the geographer's profession. Although there are major structural problems in the syllabuses of degree and bachelor's programmes, there is also the prospect of improvement and curricular reform in the future.

The aim of this work is not to exhaust the subject in question, but rather to spark new debates so that improvements can be made for the next generations of teachers to come.

REFERENCES

BLASCHKE, T.; KUX, H. **Remote sensing and advanced GIS**: new sensor systems innovative methods: 2. ed. São Paulo: Oficina de Textos, 2007.

BRAZIL. Law No. 4.024, of 20 December 1961. Establishes the Guidelines and Bases of National Education. Brasília, 1961.

BRAZIL. Presidency of the Republic. Civil House. Sub-Cabinet for Legal Affairs. Law No. 9.394, of 20 December 1996. Establishes the guidelines and bases of national education. Available at: <http://www.planalto.gov.br/ccivil_03/LEIS/l9394.htm>. [Accessed on: 6 June 2010].

BRAZIL. Secretariat for Higher Education. National Curriculum Parameters for Geography. Brasília: MEC, 1998. p.197.
DEMO, P. **Educar pela pesquisa**. Campinas: Autores Associados, 2003.

FITZ, P. R.; **Geoprocessing without complications**. São Paulo: Oficina de Textos, 2008.

FLORENZANO, T. G. **Geomorphology**: current concepts and technologies. São Paulo: Oficina de Textos, 2008.

FLORENZANO, T.G. **Satellite images for environmental studies**. São Paulo: Oficina de Textos, 2002.

II REGIONAL SYMPOSIUM ON GEOPROCESSING AND REMOTE SENSING; EDUCATION IN GEOTECHNOLOGIES: REALITIES AND CHALLENGES, 11. 2004. Aracaju. **Proceedings...** Aracaju: UFS, 2004. 1 CD-ROM.

LACOSTE, Y. **A geografia**: isso serve, em primeiro lugar, para fazer a guerra. 4. ed. Campinas: Papirus, 1997.LIMA, M.G. A pesquisa acadêmica e sua contribuição para a formação do professor de Geografia. In:PONTUSCHKA, N. N. (Org). **Geography in perspective**. 2. ed. Contexto. São Paulo: 2004. p. 115-119.

MORAIS, A. C. R. **Geografia**: pequena história crítica. 15. ed. São Paulo: Hucitec, 1995.

OLIVEIRA, A. U. Situations and Trends in Geography. In: VESENTINI, J.W. (Org). 5. ed. São Paulo: Contexto, 1994. P. 24-30.

ROSA, R. Geotechnologies in applied geography. **Revista do Departamento de Geografia INPE,** São Paulo, n. 16, p. 81-90, Sep/Oct 2005.

ROSA, R. Geotechnologies in applied geography: diffusion and access. **Revista do Departamento de Geografia INPE,** São Paulo, n. 17, p. 24-29, Sep/Oct 2005.

SUERTEGARAY; D. M. A. O atual e as tendências do ensino e da pesquisa em geografia no Brasil. **Revista do Departamento de Geografia INPE,** São Paulo, n. 16, p. 38-45, Sep/Oct, 2005.

VESENTINI, J.W. Geography teacher training - some reflections. In: PONTUSCHKA, N. N. (Org). **Geografia em perspectiva**. 2. ed. São Paulo: Contexto, 2004. p.233-235.

APPENDICES

APPENDIX A - SURVEY QUESTIONNAIRE FOR INTERVIEWS WITH TEACHERS

1. In the subject you teach, is there direct or indirect use of Geotechnology or Geoprocessing resources (Digital Cartography, Topography, Geographic Information System, GPS, Remote Sensing)?

2. Does the Geography degree course offer an adequate physical structure for the use of geotechnologies? Please explain.

3. What physical and technological resources are available for teachers to use?

4. How do you assess the importance of using geotechnologies in your subject(s) in the degree course and in the bachelor's degree course?

5. Do you feel prepared to work with geotechnologies in your subject(s) or do you believe that you lack more theoretical and/or instrumental support to use these resources in your classes? Please explain:

APPENDIX B - INTERVIEW SCRIPT PROVIDED BY THE TEACHER COORDINATOR OF THE GEOGRAPHY DEGREE COURSE

1. Why are there no subjects such as Remote Sensing and Geographic Information Systems or Geotechnologies applied to Geography on the Geography degree programme?

2. What are the reasons for not implementing these subjects in the Geography degree course?

3. In your view, would it be possible to implement the subjects of Remote Sensing and Geographic Information System in the curriculum of the Geography degree course? Why?

ANNEX

ANNEX A - TERM OF AUTHORISATION FOR THE USE OF VOICE, NAME, SOUND AND IMAGE

Federal Technological University of Paraná

Research and Postgraduate Management

Specialisation in Science and Technology Education

TERM OF AUTHORISATION FOR THE USE OF
VOICE, NAME, SOUND AND IMAGE

I would like to ask you to authorise the use of your sound and image rights for an interview to make it possible to carry out a monographic research project for the 4th Scientific and Technological Specialisation course at the Federal Technological University of Paraná.

I, single ( ) married ( )

profession: resident at Rua, nº complemento, cidade estado, bearer of ID card RG, registered with CPF/MF under n°, **I authorise** the use of my image, name, voice and sound. Therefore, as this is an expression of my will, I declare that I freely and spontaneously authorise the use described above, without anything being claimed by way of rights related to my image or any other, and I sign this authorisation in 02 copies of equal content and form.

Ponta Grossa, de de.

I want morebooks!

Buy your books fast and straightforward online - at one of world's fastest growing online book stores! Environmentally sound due to Print-on-Demand technologies.

Buy your books online at
www.morebooks.shop

Kaufen Sie Ihre Bücher schnell und unkompliziert online – auf einer der am schnellsten wachsenden Buchhandelsplattformen weltweit! Dank Print-On-Demand umwelt- und ressourcenschonend produzi ert.

Bücher schneller online kaufen
www.morebooks.shop

Printed by Books on Demand GmbH, Norderstedt / Germany